Food and You

Feeding the World with Modern Agricultural Biotechnology

A publication of the

ACSH American Council on Science and Health

Science. Not Hype.

Written by

William P. Kucewicz

Based on a book authored by

Martina Newell-McGloughlin, D.Sc.

Director, UC Systemwide Biotechnology Research and Education Program

University of California, Davis

Bruce Chassy, Ph.D.

Professor Emeritus, Department Food Science and Human Nutrition

University of Illinois at Urbana-Champaign

Gregory Conko, J.D.

Executive Director and Senior Fellow

Competitive Enterprise Institute

For more information, contact:

American Council on Science and Health
1995 Broadway, Suite 202
New York, New York 10023-5860
Tel. (212) 362-7044 • Fax (212) 362-4919
URL: http://www.acsh.org • Email: acsh@acsh.org

Publisher name: American Council on Science and Health
Title: Food and You: Feeding the World with Modern Agricultural Biotechnology
Price: $6.95
Author: William P. Kucewicz
Subject (general): Science and Health
Publication Year: 2013
Binding Type (i.e. perfect (soft) or hardcover): Perfect
ISBN: 978-0-9910055-0-5

Acknowledgements

This publication is based on the book Food and You: A Guide to Modern Agricultural Biotechnology (ISBN 978-0-9727094-8-4) by Dr. Martina Newell-McGloughlin, Dr. Bruce M. Chassy and Gregory Conko, J.D., published by the American Council on Science and Health. The following people reviewed the original book on which this summary is based:

Kent Bradford, Ph.D.
Distinguished Professor of Plant Sciences
Director, Seed Biotechnology Center
University of California, Davis

James I. Burke, Ph.D.
Professor and Chair of Crop Science
University College Dublin

Christopher J. Leaver, Ph.D.
Fellow of St John's College
University of Oxford

Alan G. McHughen, D.Phil.
CE Plant Biotechnologist
University of California, Riverside

Table of Contents

Executive Summary

Global demands on agriculture are certain to increase in tandem with increasing world population, living standards and longevity, particularly among developing countries. With limited arable land, innovative techniques will be required to improve the efficiency of the global agriculture sector to ensure an ample supply of healthful food. Biotechnology offers the most efficient, cost-effective means of raising agricultural productivity worldwide.

As long ago as 1987, an analysis published by the National Academy of Sciences (USA) examined the available literature and concluded that plants and other organisms produced using genetic engineering techniques posed no new or different risks to human health or the environment than those produced using other breeding methods. Since then, the same conclusion has been reached by a number of other respected scientific organizations. Research has shown the insertion of transgenes (i.e., genes from another source) produces less unintended DNA modification than classical plant breeding methods. The consensus of scientific opinion is that the application of genetic modification technology introduces no unique food safety or environmental impact concerns and that there is no evidence of harm from those products that have been through a regulatory approval process.

The U.S. National Research Council in 2000 determined that no difference exists between crops modified through modern molecular techniques and those modified by conventional breeding practices. The NRC emphasized that the authors were not aware of any evidence suggesting foods on the market today are unsafe to eat because of genetic modification. In fact, the scientific panel concluded that growing such crops could have environmental advantages over other crops.

The very first commercial approval of a biotech crop was granted in 1993. Biotech crops are now grown on roughly 10 percent of global cropland, even though the cultivation of biotech crops is banned in most countries. The first biotechnology products commercialized in agriculture were crops with improved agronomic traits (primarily pest and disease resistance and herbicide tolerance) whose value was unclear to consumers. Currently under development are crops with a more diverse set of new traits that can be grouped into four broad areas: 1) improvement of agronomic traits (e.g., increased yield, and resistance to climate or soil aberrancies); 2) crop plants for use as biomass feedstocks for biofuels; 3) the introduction of value-added traits, such as improved nutrition, of special importance to those populations often suffering from malnutrition; and 4) the use of plants as production factories for therapeutics and industrial products.

Two of the biggest impediments to the use of biotechnology in agriculture are governmental biosafety rules and adverse public opinion. Therefore, the actual commercialization of biotech products in the future may have less to do with technical challenges and more to do with external constraints, primarily overly stringent regulatory approval standards based on a hazard, or precautionary standard, rather than on a risk-based evaluation.

1 Introduction

In the latter half of the 20th century, major strides were made in agricultural productivity, helping to quadruple annual agricultural production worldwide. Among the factors contributing to the gains were the intensive use of chemical fertilizers, pesticides, and herbicides, the availability of new agricultural equipment, better irrigation, and selective breeding programs. Interestingly, the selective breeding of crop plants and farm animals is far from new. Agriculturalists over the millennia have employed crossbreeding, mutation selection, and the culling out of undesirable characteristics to modify animals and plants (Chrispeels and Sadava, 2003).

All of these breeding methods depend on the selection of novel traits that arise from a variety of DNA mutations. That is to say, the desired novel traits are the result of genetic changes. Thus, from a scientific perspective, the term "genetically modified organism" need not apply solely to the products of modern biotechnology, as virtually all domesticated crops and animals have been subjected to varying degrees of genetic modification. Plant and animal breeders – especially in the last century – expanded the tools of genetic manipulation beyond conventional crossbreeding to use a number of other techniques. In the case of plants, chromosome doubling and mutation breeding were achieved through the use of radiation or chemicals (Chrispeels and Sadava, 2003). And a variety of other sophisticated laboratory methods have been used to change plants in ways that do not occur in nature."

Crops developed using such methods are now common throughout the food chain. Seedless varieties of banana and watermelon, for example, were developed by tripling the number of chromosomes. Bread wheat, developed thousands of years ago, is called an allopolyploid plant because it contains six entire sets of chromosomes from three different species. Broccoflower was developed using a technique known as embryo rescue, and male sterility in cauliflower was produced by fusing together radish and cauliflower protoplasts (i.e., cells with their cell walls removed to enable the passage of DNA). Many common tomato varieties are the result of wide crosses between domesticated tomato and its wild relatives, which contain high levels of poisonous glycoalkaloid toxins. Common varieties of Asian pear and grapefruit were developed with irradiation, or mutation breeding, for fungal resistance, and the same techniques were used to modify starch in durum pasta wheat (Newell-McGloughlin, 2008). Innovations such as these have been essential for sustaining and enhancing agricultural productivity over the decades.

Historically, agriculturists and plant breeders selected improved crops based on changes that arose as a result of genetic modification of DNA (i.e., naturally occurring mutations) without any knowledge of the nature of the molecular modifications that had occurred in the DNA, or the resulting changes in the content of proteins and metabolites in newly selected varieties. Insight into the molecular changes that occur as a result of plant breeding has emerged with the introduction of high throughput DNA sequencing, coupled with improved methods for evaluating the proteome (i.e., the full complement of proteins that occur within a cell, tissue, or organism) and metabolome (i.e., the full complement of metabolites, or chemical substances, within a biological sample) of crop plants. The kinds of DNA modifications associated with classical plant breeding and transgene insertion have been assessed and compared (Parrott, 2005; Parrott et al., 2010; Weber et al., 2012).

A significant body of evidence now demonstrates that all forms of plant breeding introduce a variety of changes in DNA, ranging from point mutations and single base pair deletions and insertions to loss or acquisition of genes and changes in numbers of whole chromosomes. Of particular importance, transgene insertion (what we now call genetic modification or modern biotechnology) has been observed to produce less unintended DNA modification when compared to classical plant breeding methods. Studies also have shown that transgenic crop varieties more closely resemble their parental lines than do other varieties of the same crop with respect to their genome's protein products and metabolites (i.e., their proteomic and metabolomics profiles), and how these interact to determine biological functions. (Ricroch et al., 2011). Given the substantial and unpredictable genetic modifications in crop plants common in the human diet, the comparatively simple and more precise modifications performed with recombinant DNA techniques appear to be unique only in the breeder's improved ability to control the results.

2 What is Biotechnology?

In the simplest and broadest sense, biotechnology (also known as modern biotechnology, or new biotechnology) is a series of enabling technologies that involve the manipulation of living organisms, or their sub-cellular components, to develop useful products and processes. The capacity to manipulate the genetic makeup of living organisms with precision has become one of the cornerstones of modern biotechnology. It enables developers to enhance the ability of an organism to produce a particular chemical product (e.g., penicillin from a fungus), to prevent it from producing a product (e.g., ethylene in plant cells), or to enable it to produce an entirely new product (e.g., chymosin in microorganisms).

Biotechnology has introduced a new dimension to selective breeding, offering a number of efficient, cost-effective means to improve crops. Such transgenic crops are often referred to as "genetically modified organisms," or GMOs. We will refer to them as "genetically engineered," or GE, organisms. Biotechnology has the potential to improve both the qualitative and quantitative aspects of food, feed, and fiber production. In time, it also may reduce agriculture's dependency on chemicals by transitioning to new biological solutions and thus moderate raw-material costs all in an environmentally sustainable manner.

It is further possible to enhance the nutritional content, texture, color, flavor, growing season (i.e., time to flowering), yield, disease and pest resistance, and other properties of production crops. Transgenic techniques can be applied to farm animals to improve their growth, fitness and other qualities. Enzymes produced using recombinant DNA methods (in microorganisms/bacteria and yeasts/fungi, etc.) are used to make cheese, keep bread fresh, produce fruit juices and wines, and treat fabric for blue jeans and other denim clothing. Other recombinant DNA enzymes are used in laundry detergents. Recombinant microorganisms can be engineered to improve environmental quality, too. Bioprocessing using engineered microbes offers new ways to treat waste. And, in a process known as bioremediation, naturally occurring microorganisms are being used to treat organic and inorganic contaminants in soil, groundwater, and air.

Proliferation of Biotech Products

3

The very first commercial approval of a biotech crop — Celgene's Flavr-Savr tomato — was granted in 1993. Biotech crops are now grown on roughly 10 percent of global cropland (James, 2013), even though the cultivation of biotech crops is banned in most countries based upon political/precautionary reasons rather than scientific evidence. The first biotechnology products commercialized in agriculture were crops with improved agronomic traits, primarily pest and disease resistance and herbicide tolerance. And the benefits of biotechnology to agriculture are bound to grow in importance as the world's population expands from the current 7 billion to a forecast 9 billion by 2050. According to some estimates, agricultural production over the next 25 years will have to double just to keep pace with rising demand.

Modifications of crop plants can be organized into two broad, non-exclusive categories: those that benefit the producer through introduction of such properties as improved insect, weed, and disease management, and lower input costs; and those that benefit the consumer more directly, with increased nutritional value, flavor, or other desirable product attributes. Many plants also deliver benefits for the environment, such as reducing insecticide use and hastening an ongoing shift to conservation tillage practices. Modifications that increase total crop yield or protect a crop from either biotic stress (i.e., damage by predators, such as insects, weeds, or disease agents, including viruses, fungi, and bacteria) or abiotic stress (i.e., damage from other causes, such as drought, flooding, cold, heat, salination, or poor soil) primarily benefit the producer and are often called "input traits." Researchers have only begun to tap the potential of biotechnology to produce varieties of plants that confer direct advantages for consumers in their consumption. Varieties modified to have greater appeal to consumers are said to have enhanced "output traits." The majority of biotech crops in commercial use today fall into the input category.

Among the plant varieties currently marketed, the most common traits are insect resistance, herbicide tolerance, and virus resistance. The pest-resistance trait was added by inserting a gene from the common soil bacterium Bacillus thuringiensis (Bt), which produces an insoluble crystalline protein that adheres to and degrades the alkaline stomach of only one or a very few species of insect larvae.

Tolerance to a different herbicides is another sought-after trait. Weeds compete with crop plants for sunlight, water, and soil nutrients, and if not eliminated, they can lead to significant yield losses. Consequently, effective weed management is essential to production-scale agriculture. With herbicide-tolerant crops, growers can spray a broad-spectrum herbicide on their fields, effectively managing all or most weed species, while leaving the crop plants unharmed. The last major class of biotech traits now on the market is virus resistance.

The plant species most widely adopted with gene-splicing methods (that is, the process in which fragments of DNA from one or more different organisms are combined to form recombinant DNA and are made to function within the cells of a host organism) are the commodities corn, cotton, soy, and canola. U.S. farmers grow each of these crops and also have planted a significant number of acres with biotech varieties of sugar beet and alfalfa, while a far smaller number of acres have been planted with biotech squash, papaya, and rice. Rice is the principal staple for much of the world, and corn is the largest animal feed source, so rising productivity in those two crops globally will have important impacts on long-term food security.

The United States has the largest number of approved and commercially planted biotech varieties. The primary federal body in charge of regulating biotech plants, the U.S. Department of Agriculture, has approved (or, in agency parlance, "deregulated") more than 90 "transformation events" (i.e., organisms resulting from transgene insertions) of 16 plant species for commercial-scale cultivation, though many of these products, while legal to grow and sell, are not commercially available. While scores of countries still forbid the planting of any GE crop (Paarlberg 2001, 2010), there is movement toward wider acceptance. The first GE crop to be released for commercial cultivation in India was Bt cotton, for instance. China, too, has begun to set the pace for new approvals, becoming the first major rice-producing country to approve a GE rice variety and granting initial approval for a maize variety engineered to reduce the amount of phosphate in the waste from corn-fed livestock.

Where GE crops have become available, many farmers have eagerly planted the new varieties. By some estimates, biotech crops have been the most rapidly adopted agricultural technology in history. U.S. farmers grow the largest number of acres (over 150 million) planted with biotech varieties, accounting for about 43 percent of the total acreage worldwide. The U.S. is followed in terms of acreage by Brazil (67m), Argentina (53m), India (23m), and Canada (23m). Twenty-nine countries now plant genetically-engineered crops (James 2013). Twenty of the 29 nations are less developed countries (LDCs), and 90 percent of the farmers, or about 15.6 million, are in LDCs (James, 2012). The most recent countries to join this group include Uruguay, Paraguay, Bolivia, Egypt, Burkina Faso, Pakistan, and Myanmar (Burma). In 2010, Germany resumed the legal planting of biotech crops after withdrawing authorization several years earlier; the European Union, however, has approved only a small number of GE varieties for import, and even fewer are actually allowed to be grown there. An additional 31 countries have permitted pre-commercial field trials of biotech crop varieties, or have approved some harvested biotech plants to be imported for use as food and livestock feed. Nonetheless, a majority of countries continue to prohibit transgenic crops and GE food imports.

4

A Distinction Without a Difference

All new varieties of crops are the result of genetic modification regardless of the technology used for their development. To date, new crop varieties have been almost without exception safe to plant and safe to consume. The small number of documented cases in which a new variety was found to be unsafe for consumers were all the products of classical breeding methods (NRC 2004). Nevertheless, new varieties have proven so comparatively safe that non-biotech ones are released to farmers with essentially no oversight by regulators and, with very few exceptions, no requirements for safety testing. Crops produced using modern biotechnology are, however, all subject to special regulation with associated significant cost implications.

Aspects of the regulatory framework in every country that permits the commercial use of biotech crops, or food and animal feeds derived from them, are premised on the belief that unique risks arise from the transformation process itself. Each time a gene is introduced into a plant, the resulting organism (or transformation event) is treated as a unique product for the purposes of regulation. Even if copies of a single gene encoding the same protein are inserted into different plants of the same species, each resulting transformation event must be tested and approved separately. **There is no evidence, however, that the uncertainties associated with transgene insertion are any greater than those that occur with other forms of genetic modification, such as the random genetic changes that result from mutation breeding.**

Critics of transgenic crops claim the use of modern biotechnology in agriculture is intrinsically unsafe. But such criticism is based on a major misunderstanding. The critics accept as a given the safety of new crops developed in the customary manner. Yet traditional selective breeding methods depend on novel genetic traits that arise from DNA mutations. Plants created by these conventional phenotypic selection techniques undergo no formal food or environmental safety evaluation other than normal agricultural variety testing prior to introduction into the environment or marketplace. This is not to suggest that classical breeding methods are inherently unsafe. What it does suggest is that the contrasting regulatory treatment of these two classes of plants is arbitrary, merely adding a needless, burdensome obstacle to innovation while adding tremendously to their costs.

As long ago as 1987, an analysis published by the National Academy of Sciences (USA) examined the available research and concluded that plants and other organisms produced using genetic engineering techniques pose no new or different risks to human health or the environment than those produced using other breeding methods (NAS, 1987). Since that time, the National Academies, the European Union, and the governments of a number of other countries have reviewed the scientific literature and reached the same conclusion.

Transgenic crops produced using the new biotechnology are nonetheless regulated by governments and may not be released to farmers or consumers until they have successfully passed a rigorous pre-market safety assessment (Kok and Kuiper, 2003; König et al., 2004; Codex, 2003). On a case-by-case basis, the safety assessor seeks to determine if the new trait introduced into a crop is cause for safety concerns. In principle, the focus of regulators is on the safety of the new trait and not on the fact that genetic engineering has been used to introduce the new trait. Yet, paradoxically, crops developed using less precise and more disruptive methods of breeding may be released without any pre-market regulatory review.

It is commonly believed that transgenic crops should be regulated because they express novel traits not normally associated with that crop, typically not part of the human or animal diet. When a genuinely novel substance (e.g., a new protein or other phytochemical) is introduced into a plant, this does merit special testing to ensure safety. But most of the traits introduced into biotech crops can also be introduced with various classical breeding methods. Moreover, classical breeding methods, such as interspecies and intergenera "wide cross" hybridization, frequently introduced new genes and gene products into the human diet. Thus, not all biotech plant varieties contain genes or proteins new to the food supply, neither is the introduction of novel substances unique to transgenic breeding methods.

It bears repeating that the only scientific justification for pre-market safety assessment for any new plant variety is to establish the safety of any newly introduced substances. It is an unfortunate reality that pre-market safety assessment has become an endless search for unintended effects. Consider what happened in the Philippines. The biotech crop at issue was none other than the flagship of improved nutritional varieties: beta carotene-enhanced rice, commonly referred to as "golden rice." Authorities in the Philippines had under consideration since the late 1990s an application to plant golden rice. Yet, despite numerous risk assessments, the modified crop did not win governmental approval until February 2013. The developer, Ingo Potrykus — who with his colleagues was working for a publicly-funded research institute — lays the blame largely on the regulatory process, which he considers excessive, and he pointedly observes that similar legal requirements in many countries are preventing genetically engineered crops from saving millions from starvation and malnutrition.

No Evidence of Harm

The consensus of scientific opinion and evidence is that the application of GE technology introduces no unique food or feed safety concerns or environmental impacts and that there is no evidence of harm from those products that have been through a regulatory approval process. This conclusion has been reached by numerous national and international bodies, including the Food and Agriculture Organization of the United Nations, the World Health Organization, the Organization for Economic Cooperation and Development, the European Commission, the French Academy of Sciences, the U.S. National Research Council of the National Academy of Sciences, the Royal Society of London, and the Society of Toxicology.

Take the U.S. National Research Council, for instance. In its report "Genetically Modified Pest-Protected Plants: Science and Regulation" (NRC, 2000), it determined that no difference exists between crops modified through modern molecular techniques and those modified by conventional breeding practices. The NRC emphasized that the authors were not aware of any evidence suggesting foods on the market today are unsafe to eat because of genetic modification. In fact, the scientific panel concluded that growing such crops could have environmental advantages over other crops. In a 2003 position paper, the Society of Toxicology (SOT, 2003) corroborated this finding and noted that there is no reason to suppose that the process of food production through biotechnology leads to risks of a different nature than those already familiar to toxicologists or to risks generated by conventional breeding practices for plant, animal or microbial improvement. It is therefore important to recognize that it is the food product itself, rather than the process through which it is made, that should be the focus of attention in assessing safety.

Similarly, a European Commission report (EU, 2001, 2008) that summarized biosafety research of 400 scientific teams from all parts of Europe conducted over 15 years stated that research on GE plants and derived products so far developed and marketed, following usual risk assessment procedures, has not shown any new risks to human health or the environment beyond the usual uncertainties of conventional plant breeding. Indeed, the use of more precise technology and the close regulatory scrutiny probably make GE plants even safer than conventional plants and foods. More recently, EU-funded research from 130 projects involving 500 independent research groups over 25 years concluded, "There is, as of today, no scientific evidence associating GMOs with higher risks for the environment or for food and feed safety than conventional plants and organisms" (Europa Press Release, 2010). What is more, the lack of any credible reports of adverse effects resulting from the production and consumption of GE crops grown on more than 235 million cumulative hectares over the last seven years further supports these safety conclusions.

6 Biosafety Testing

In contrast to traditionally bred crops, a rigorous safety-testing paradigm has been developed and implemented for GM crops (Cockburn, 2002; Kok and Kuiper, 2003; König et al., 2004). The science-based process focuses on a classical evaluation of the toxic potential of the introduced novel gene, its gene product, and the wholesomeness for human consumption of the GE crop. In addition, detailed consideration is given to the history and safe use of the parent crop, as well as that of the gene donor(s). The overall safety evaluation is conducted using a process known as "substantial equivalence," a model that is entrenched in all international crop biotechnology guidelines (Kok and Kuiper, 2003; Codex, 2003). The paradigm provides the framework for a comparative approach to identify the similarities and differences between the GE product and an appropriate comparator that has a known history of safe use. The information is used to reach a conclusion about whether food or feed derived from the GE crop is as safe as food or feed derived from its traditional counterpart or the appropriate comparator.

Substantial equivalence is only one in an array of principles employed in the international consensus approach to safety assessment of transgenic crops (Chassy et al., 2004; Chassy et al., 2008; Ricroch et al., 2011). Other governing principles are listed below:

- **Potential gene transfer:** Where there is a possibility that selective advantage may be given to an undesirable trait from a food safety perspective, this should be assessed. An example is in the highly unlikely event of a gene coding for a plant-made pharmaceutical being transferred to a food crop (e.g., corn). Where there is a possibility that the introduced gene(s) may be transferred to other crops, the potential environmental impact of the introduced gene and any conferred trait must be assessed.

- **Potential allergenicity:** Since most food allergens are proteins, the potential allergenicity of newly expressed proteins in food must be considered. The starting point of this decision-tree approach, first introduced in 1996, is the known allergenic properties of the donor (gene-source) organism. Other recurrent items in this approach are structural similarities between the introduced protein and allergenic proteins, digestibility of the newly introduced protein(s), and if needed, immunological assays known as sera-binding tests.

- **Potential toxicity:** Some proteins are known to be toxic (e.g., enterotoxins from pathogenic bacteria and lectins from plants). Tests for toxicity include comparisons of amino acid sequences of any newly expressed protein(s) with the amino acid sequences of known toxins, as well as rodent toxicity tests with acute administration of the proteins.

- **Unintended effects:** Interactions of the inserted DNA sequence with the plant genome are possible sources of unintended effects. Another source might be the introduced trait unexpectedly altering plant metabolism. The process of product development that selects a single commercial product from hundreds to thousands of initial transformation events eliminates the vast majority of situations that might have resulted in unintended changes. The selected commercial product candidate then undergoes additional detailed analyses to further screen for unwanted effects.
- **Long-term effects:** It is acknowledged that the pre-market safety assessment should be rigorous to exclude potentially adverse effects of consumption of foods or feeds derived from GE crops. Nevertheless, some have insisted that such foods should also be monitored for long-term effects by post-market surveillance. No international consensus exists as to whether such surveillance studies are technically possible without a testable hypothesis in order to provide meaningful information regarding safety, and a GE crop with a testable safety concern would most likely not pass regulatory review.

GE Food Labeling

The question of whether foods derived from organisms modified with recombinant DNA techniques should be specially labeled has received a great deal of attention. The U.S. Food and Drug Administration's (FDA, 1992) approach to the labeling of foods, including those genetically engineered or otherwise novel, is that the label must be accurate and "material." Agency officials recognize that any breeding method could impart a change that makes food less safe or nutritious than its conventional counterpart but that the process of recombinant DNA (rDNA) modification, in which a DNA molecule is formed by joining DNA segments from different sources, is not inherently risky. Accordingly, special labeling is required "if a food derived from a new plant variety differs from its traditional counterpart such that the common or usual name no longer applies, or if a safety or usage issue exists to which consumers must be alerted."

Such changes include the introduction of a toxin, anti-nutrient, or allergen into a food product in which consumers would not ordinarily expect to find it (e.g., an allergenic protein from nuts in corn), the elevation of an endogenous substance to potentially harmful levels (e.g., a significant increase in potato or tomato glycoalkaloids), or a significant change in the level of dietary nutrients in a food (e.g., oranges with abnormally low levels of vitamin C). Other material changes that must be labeled include those that relate to the storage, preparation, or usage characteristics of a food, such as a change affecting the length of time or manner in which kidney beans must be soaked and cooked before eating, or the safe shelf-life of various food products. Even a change in organoleptic (sensory) characteristics of a food from what consumers would normally expect, including the taste, smell, or mouth feel of a food, is considered material and must be labeled. Importantly, the FDA's policy stipulates that the altered characteristic itself must be specified on the label, not the breeding method used to impart the change.

The FDA also emphasizes that no pre-market review or approval is required unless characteristics of the biotech food explicitly raise safety issues. Indeed, the FDA cannot require the labeling to make mention of the genetic method used in the development of a new plant variety. Obviously, many of the novel nutritionally enhanced foods expected on the market in the next few years will be labeled, as they will differ from their traditional counterparts, and in most instances marketers will want to proclaim the new product's enhanced nutritional value.

8 Barriers to Biotechnology

Biotech crop developers, seed breeders, and farmers face a number of hurdles when introducing new varieties. Although transgenic crops are grown in 29 countries, the technology has met stiff resistance from some consumers, producers, non-governmental organizations (NGOs), and regulators. Many countries ban both the cultivation of GE crops and the import of food or animal feeds derived from them. Yet, even in the countries where GE crops are grown, such as the United States and Canada, the vast majority of production is limited to cotton and commodity grains (e.g., corn, canola, and soy) that are primarily fed to livestock or consumed by humans only after processing. Despite the significant economic benefit reaped by producers of GE commodity crops, very few GE varieties of whole fruits or vegetables are grown commercially. The explanation for this phenomenon is complex and multifaceted, but consumer attitudes, food industry ambivalence, production costs, regulatory impediments, and market access all play a role.

U.S. consumer attitudes tend to be mixed on food biotechnology. An International Food Information Council Survey in 2012 found that 38 percent of respondents held a favorable opinion of using biotechnology to produce food, while 20 percent had a negative opinion. Still, 77 percent said they would be likely to purchase foods bioengineered to require less pesticide use, and 71 percent said they would buy foods made with cooking oils modified to have a healthier fat content. Outside the U.S., public attitudes vary widely, with consumers in Europe expressing the most significant opposition to GE products. Perhaps more important, a general lack of knowledge about GE foods means few consumers are aware of the benefits of these products.

A bigger problem than consumer resistance is the rejection of biotech foods by producers and retailers. A small but important segment of the public holds very passionate anti-biotechnology attitudes. In response, many packaged-food companies and food retailers have been reluctant to embrace GE products. With anti-biotechnology campaigners eager to protest against supermarket chains and food processing companies who use bioengineered ingredients, it is understandable that few firms are willing to put their hard-earned reputations at risk. And the bigger the companies, the less willing they seem to be to use biotechnology (Kalaitzandonakes and Bijman, 2003).

Regulations pose even greater difficulties. In the U.S., regulatory compliance adds at least $1 million to the cost of developing a GE variety for each transformation event (Redenbaugh and McHughen, 2004). For crops intended for international commerce, the regulatory costs in key producing and importing countries have been estimated to range from $6 million to $15 million (Kalaitzandonakes et al., 2007) and as high as $35 million (McDougall, 2011). Unjustifiably burdensome rules add little to environmental and human health protection and, arguably, do more harm than good, especially in less developed countries. Complicating matters further is the highly charged political environment in which biosafety regulatory decisions are made.

9 Conclusion

The United Nations' Food and Agriculture Organization (FAO) estimates that about one billion people worldwide suffer from under-nutrition, to which insufficient protein in the diet is a significant contributing factor (FAO, 2012). Protein-energy malnutrition (PEM) is the most lethal form of malnutrition and affects every fourth child worldwide (WHO, 2006). Biotech crops could potentially do much to relieve the problem of under-nutrition. Plant-based products comprise the vast majority of human food intake, irrespective of location or financial status (Mathers, 2006). Indeed, in some cultures, either by design or default, plant-based nutrition comprises almost 100 percent of the diet. Given this fact, it can be deduced that significant nutritional improvement could be achieved via modifications of staple crops to deliver higher micronutrient levels.

There are no alternative technologies available to plant breeders with which needed phenotypes — new improved varieties — can be created, none which can overcome the physiological and environmental limitations of global agriculture to produce sufficient food, feed, fuel, and fiber on the available arable land to meet increasing demand (i.e., sustainable intensification). The scientific hurdles to achieving these goals are not trivial – particularly as researchers strive to modify qualitative, as opposed to quantitative, traits and alter intricate metabolic pathways and networks, as opposed to single genes. However, the tools now coming on line in the fields of genomics, proteomics, and the like are bound to offer solutions.

Non-technical factors pose different challenges. Among these are intellectual property restrictions, liability concerns, prohibitive biosafety rules, and public acceptance. The last two in many ways are the most insidious of limitations on biotechnology as they have little basis in fact and thus are difficult to refute effectively. It is easier to appeal to emotion and sell fear than it is to present a reasoned and judicious scientific rationale on which to base risk analysis. Looking forward, the actual commercialization of biotech products may have less to do with technical limitations and more to do with external constraints, primarily the process of regulatory approval.

Sources

Abbadi, A.D.F. et al. (2004) Biosynthesis of very-long-chain polyunsaturated fatty acids in transgenic oilseeds: constraints on their accumulation. Plant Cell 16, 2734–2748

Agbios. Agbios 2008 GM crop database. Available from: http://www.agbios.com/dbase.php?action=ShowForm.

Agius, F. et al. (2003) Engineering increased vitamin C levels in plants by overexpression of a D-galacturonic acid reductase. Nat. Biotechnol. 21, 177–181

Amyris (2013) Amyris Biofuel research http://www.amyris.com/Innovation/155/BreakthroughScience

Anai, T. et al. (2003) Improvement of rice (Oryza sativa L.) seed oil quality through introduction of a soybean microsomal omega-3 fatty acid desaturase gene. Plant Cell Rep. 21, 988–992

Arcadia Biosciences and Bioriginal Food and Science Corp. Enter Strategic Alliance to Market High GLA Safflower Oil. Business Wire, Feb. 22, 2008. Available at: http://www.arcadiabio.com/news/press-release/arcadia-biosciences-and-bioriginal-food-and-science-corp-enter-strategic-alliance and http://findarticles.com/p/articles/mi_m0EIN/is_2008_Feb_22/ai_n24320185/

Arcadia, NUE (2013) Nitrogen Use efficiency, http://www.arcadiabio.com/nitrogen

Atanassov, A.B.A. et al. (2004) To reach the poor – results from the ISNAR-IFPRI Next Harvest study on genetically modified crops, public research, and policy implications. EPTD Discussion Paper – Environment and Production Technology Division International Food Policy Research Institute

Austin-Phillips, S.B.E. et al. (1999) Production of Industrial and Animal Feed Enzymes in Transgenic Alfalfa. Available from: http://www.molecularfarming.com/nonmedical.html

Brinch-Pedersen, H. et al. (2000) Generation of transgenic wheat (Triticum aestivum L.) for constitutive accumulation of an Aspergillus phytase. Mol. Breed. 6, 195–206

Brookes, G. & Barfoot P. (2012a) The income and production effects of biotech crops globally 1996–2010. GM Crops and Food: Biotechnology in Agriculture and the Food Chain 3:265 - 272; PMID: 22750951; http://dx.doi.org10.4161/gmcr.20097

Bruening, G. & Lyons, J.M. (2000) The case of the FLAVR SAVR tomato. California Agriculture 54(4):6-7. DOI: 10.3733/ca.v054n04p6. July-August 2000

Cahoon, E.B. et al. (2003) Metabolic redesign of vitamin E biosynthesis in plants for tocotrienol production and increased antioxidant content. Nat. Biotechnol. 21, 1082–1087

Caimi, P.G. et al. (1996) Fructan accumulation and sucrose metabolism in transgenic maize endosperm expressing a Bacillus amyloliquefaciens SacB gene. Plant Physiol. 110, 355–363

Chakraborty, S. et al. (2000) Increased nutritive value of transgenic potato by expressing a nonallergenic seed albumin gene from Amaranthus hypochondriacus. Proc. Natl. Acad. Sci. U. S. A. 97, 3724–3729

Chapman, K.D.A.-B.S. et al. (2001) Transgenic cotton plants with increased seed oleic acid content. J. Am. Oil Chem. Soc. 78, 941–947

Chassy, B. et al. (2004) Nutritional and safety assessments of foods and feeds nutritionally improved through biotechnology – Prepared by a task force of the ILSI International Food Biotechnology Committee. Comprehensive Reviews in Food Science and Food Safety 3, 35-104

Chassy, B. et al. (2008) Recent developments in the safety and nutritional assessment of nutritionally improved foods and feeds. Comp. Rev. Food Sci. Food Safety 7, 50-113

Chen, Z. et al. (2003) Increasing vitamin C content of plants through enhanced ascorbate recycling. Proc. Natl. Acad. Sci. U. S. A. 100, 3525–3530

Chiang, C.Y.F. et al. (2005) Expression of a bi-functional and thermostable amylopullulanase in transgenic rice seeds leads to autohydrolysis and altered composition of starch. Mol. Breed. 15, 125–143

Choudhary, B. & Gaur, K. (2008) The Development and Regulation of BT Brinjal in India (Eggplant/Aubergine). ISAAA38

Choudhary, B. & Gaur, K. (2010) BT Cotton in India: A Country Profile ISAAA

Cockburn, A. (2002) Assuring the safety of genetically modified (GM) foods: the importance of an holistic, integrative approach. Journal of Biotechnology 98, 79-106

Codex Alimentarius (2003) Codex Principles and Guidelines on Foods derived from Biotechnology. http://ftp.fao.org/codex/standard/en/CodexTextsBiotechFoods.pdf

Crispeels, M.J. & Sadava, D.E. (2003) Plants, Genes, and Crop Biotechnology, 2nd Ed., Jones and Bartlett Publishers

Cromwell, G.L. et al. (1967) Nutritional value of opaque-2 corn for swine. J. Anim. Sci. 26, 1325–1331

Del Vecchio, A. (1996) High-laurate canola. How Calgene's program began, where it's headed. INFORM – Int. News Fats Oils Relat. Mater. 7:, 230–243

Denbow, D.M. et al. (1998) Soybeans transformed with a fungal phytase gene improve phosphorus availability for broilers. Poult. Sci. 77, 878–881

Diaz de la Garza, R. et al. (2004) Folate biofortification in tomatoes by engineering the pteridine branch of folate synthesis. Proc. Natl. Acad. Sci. U. S. A. 101, 13720–13725

Dinkins, R.D.R.M. et al. (2001) Increased sulfur amino acids in soybean plants overexpressing the maize 15 kDa zein protein. In Vitro Cell. Dev. Biol. Plant 37, 742–747

Dodo, H.W., Konan K.N., Chen F.C., Egnin M., Viquez O.M. (2008) Alleviating peanut allergy using genetic engineering: the silencing of the immunodominant allergen Ara h 2 leads to its significant reduction and a decrease in peanut allergenicity. Plant Biotechnol J. 135-45

Drakakaki, G. et al. (2005) Endosperm-specific co-expression of recombinant soybean ferritin and Aspergillus phytase in maize results in significant increases in the levels of bioavailable iron. Plant Mol. Biol. 59, 869–880

Ducreux, L.J. et al. (2005) Metabolic engineering of high carotenoid potato tubers containing enhanced levels of beta-carotene and lutein. J. Exp. Bot. 56, 81–89

Enfissi, E.M. et al. (2005) Metabolic engineering of the mevalonate and nonmevalonate isopentenyl diphosphate-forming pathways for the production of health-promoting isoprenoids in tomato. Plant Biotechnol. J. 3, 17–27

Europa Press Release (2010) Commission publishes compendium of results of EU-funded research on genetically modified crops http://europa.eu/rapid/press-release_IP-10-1688_en.htm Accessed March 17, 2013

Falco, S.C. et al. (1995) Transgenic canola and soybean seeds with increased lysine. Biotechnology (N. Y.) 13, 577–582

FAO (2013) at: http://www.fao.org/publications/sofi/en/

Fraser, P.D.R.S. et al. (2001) Elevation of carotenoids in tomato by genetic manipulation. J. Sci. Food Agric. 81, 822–827

Froman, B. (2002) Genetic modification of oils for improved health benefits: production of long chain omega-3 fatty acids in plants. Abstr. Papers Am. Chem. Soc. 223 U35-U

Galili, G. et al. (2002) Genetic, molecular, and genomic approaches to improve the value of plant foods and feeds. Crit. Rev. Plant Sci. 21, 167–204

Gan, S. & Amasino, R.M. (1997) Making sense of senescence: molecular genetic regulation of leaf senescence. Plant Physiol 113:313–319

Giovinazzo, G. et al. (2005) Antioxidant metabolite profiles in tomato fruit constitutively expressing the grapevine stilbene synthase gene. Plant Biotechnol. J. 3, 57–69

Glei, M. et al. (2006) Both wheat (Triticum aestivum) bran arabinoxylans and gut flora-mediated fermentation products protect human colon cells from genotoxic activities of 4-hydroxynonenal and hydrogen peroxide. J. Agric. Food Chem. 54, 2088–2095

Gonsalves, D., Tripathi S., Carr J.B., Suzuki J.Y. (2010) Papaya Ringspot virus. The Plant Health Instructor. DOI: 10.1094/PHI-I-2010-1004-01

Gonzali, S. et al. (2009) Purple as a tomato: towards high anthocyanin tomatoes. Trends Plant Sci. 14, 237–241

Gosal, S.S., Wani S.H., Kang, M.S. (2009) Biotechnology and drought tolerance. J. Crop Improvement, 23: 19-54

Goto, F. et al. (2000) Iron accumulation and enhanced growth in transgenic lettuce plants expressing the iron-binding protein ferritin. Theor. Appl. Genet. 100, 658–664

Han, G. (2009) Origin Agritech Announces Final Approval of World's First Genetically Modified Phytase Corn. Available from: http://www.allbusiness.com/science-technology/biology-biotechnology-genetic/13453842-1.html

Hartwig, E. et al. (1997) Seed protein and its relationship to soluble sugars in soybeans. Crop Sci. 37, 770–773

Hellwege, E.M. et al. (1997) Transgenic potato tubers accumulate high levels of 1-kestose and nystose: functional identification of a sucrose 1- fructosyltransferase of artichoke (Cynara scolymus) blossom discs. Plant J. 12, 1057–1065

Hellwege, E.M. et al. (2000) Transgenic potato (Solanum tuberosum) tubers synthesize the full spectrum of inulin molecules naturally occurring in globe artichoke (Cynara scolymus) roots. Proc. Natl. Acad. Sci. U. S. A. 97, 8699–8704

Hipskind, J.D. & Paiva, N.L. (2000) Constitutive accumulation of a resveratrol-glucoside in transgenic alfalfa increases resistance to Phoma medicaginis. Mol. Plant Microbe Interact. 13, 551–562

Jalani, B.S.C.S. et al. (1997) Improvement of palm oil through breeding and biotechnology. J. Am. Oil Chem. Soc. 74, 1451–1455

James C. (2013) Global Status of Commercialized Biotech/GM Crops: 2012 ISAAA, Brief No. 44. Ithaca, N.Y.: International Service for the Acquisition of Agri-biotech Applications. Available from: http://www.isaaa.org/. Accessed 2012 Jan 15

James, M.J. et al. (2003) Metabolism of stearidonic acid in human subjects: comparison with the metabolism of other n-3 fatty acids. Am. J. Clin. Nutr. 77, 1140–1145

Kalaitzandonakes, N. et al. (2007) Compliance costs for regulatory approval of new biotech crops. Nature Biotech. 25, 509-511

Kalaitzandonakes, N. & Bijman, J. (2003) Who is driving biotechnology acceptance? Nature Biotechnology 21, 366-369

Katsube, T. et al. (1999) Accumulation of soybean glycinin and its assembly with the glutelins in rice. Plant Physiol. 120, 1063–1074

Kinney, A.J. (1998) Designer Oils: The High Oleic Acid Soybean (Roller, S.H.S., ed.), Blackie Academic and Professional

Knothe, Gerhard (2007) The Biodiesel Handbook, Chaper 2 – The History of Vegetable Oil Based Diesel Fuels, by Gerhard Knothe, ISBN 978-1-893997-79-0

Kobayashi, S.D.C. et al. (2000) Kiwifruits (Actinidia deliciosa) transformed with a Vitis stilbene synthase gene produce piceid (resveratrol-glucoside). Plant Cell Rep. 19, 904–910

Kok, E.J. & Kuiper, H.A. (2003) Comparative safety assessment for biotech crops. Trends in Biotechnology 21, 439-444

Konig, A. et al. (2004) Assessment of the safety of foods derived from genetically modified (GM) crops. Food and Chemical Toxicology 42, 1047-1088

Lai, J.S. (2002) Increasing maize seed methionine by mRNA stability. Plant J. 30, 395–402

Li, L.L.S. et al. (2001) Increase of sulphur-containing amino acids in transgenic potato with 10 ku zein gene from maize. Chin. Sci. Bull. 46, 482–484

Liu, Q. et al. (2002) High-oleic and high-stearic cottonseed oils: nutritionally improved cooking oils developed using gene silencing. J. Am. Coll. Nutr. 21 (3 Suppl.), 205S–211S

Lucca, P. et al. (2002) Fighting iron deficiency anemia with iron-rich rice. J. Am. Coll. Nutr. 21 (3 Suppl.), 184S–190S

Luciani, G. & Wofford, D.S. (2005) Over-Expression of a Soybean Vegetative Storage Protein Gene in Bahiagrass (Paspalum notatum var. Flugge). Available from: http://www.ufgi.ufl.edu/Symposium/FG2005program_rev.pdf

Lukaszewicz, M. et al. (2004) Antioxidant capacity manipulation in transgenic potato tuber by changes in phenolic compounds content. J. Agric. Food Chem. 52, 1526–1533

Mathers, J.C. (2006) Plant foods for human health: research challenges. Proc. Nutr. Soc. 65, 198-203

McCabe, M.S., Garratt L.C., Schepers F., Jordi W.J.R.M., Stoopen G.M., Davelaar E., van Rhijn H.A., Power J.B., Davey M.R. (2001) Effects of PSAG12-IPT Gene Expression on Development and Senescence in Transgenic Lettuce Plant Physiology October 2001 vol. 127 no. 2, 505-516

McDougall, P. (2011) The cost and time involved in the discovery, development and authorization of a new plant biotechnology derived trait, A Consultancy Study for Crop Life International

Muir, S.R. et al. (2001) Overexpression of petunia chalcone isomerase in tomato results in fruit containing increased levels of flavonols. Nat. Biotechnol. 19, 470–474

Naqvi, S. et al. (2009) Transgenic multivitamin corn through biofortification of endosperm with three vitamins representing three distinct metabolic pathways. Proc. Nat. Acad. Sci. U. S. A. 106, 7762–7767

Niggeweg, R. et al. (2004) Engineering plants with increased levels of the antioxidant chlorogenic acid. Nat. Biotechnol. 22, 746–754

Novozymes Biofuels (2013) http://www.bioenergy.novozymes.com/en/the-basics/Pages/default.aspx

O'Quinn, P.R. et al. (2000) Nutritional value of a genetically improved highlysine, high-oil corn for young pigs. J. Anim. Sci. 78, 2144–2149

Parrot, W. (2005) The Nature of Change: Towards Sensible Regulation of Transgenic Crops Based on Lessons from Plant Breeding, Biotechnology and Genomics. National Agricultural Biotechnology Council Report 17, Agricultural Biotechnology: Beyond Food and Energy to Health and the Environment, 209-220

Parveez, G. (2003) Novel products from transgenic oil palm. AgBiotechNet 5, 1–8

Prakash, C.S. & Jaynes, J. (2000) Increasing the protein content in sweet potato using a synthetic storage protein gene. Abstr. Papers Am. Chem. Soc. 219, U36

Rapp, W. (2002) Development of soybeans with improved amino acid composition. 93rd AOCS Annual Meeting & Expo pp. 79–86, American Oil Chemists' Society Press

Reddy, A.S. & Thomas, T.L. (1996) Expression of a cyanobacterial delta 6- desaturase gene results in gamma-linolenic acid production in transgenic plants. Nat. Biotechnol. 14, 639–642

Redenbaugh, K. & McHughen, A. (2004) Regulatory Challenges Reduce Opportunities for Horticultural Crops. California Agriculture 58, 106-115

Ricroch, A.E. et al. (2011) Evaluation of Genetically Engineered Crops Using Transcriptomic, Proteomic, and Metabolomic Profiling Techniques. Plant Physiology 155, 1752-1761

Rocheford, T.R. et al. (2002) Enhancement of vitamin E levels in corn. J. Am. Coll. Nutr. 21 (3 Suppl.), 191S–198S

Roesler, K. et al. (1997) Targeting of the Arabidopsis homomeric acetyl-coenzyme A carboxylase to plastids of rapeseeds. Plant Physiol. 113, 75–81

Rosati, C. et al. (2000) Metabolic engineering of beta carotene and lycopene content in tomato fruit. Plant J. 24, 413–419

Schwall, G.P. et al. (2000) Production of very-high-amylose potato starch by inhibition of SBE A and B. Nat. Biotechnol. 18, 551–554

Scott, S.A., Davey M.P., Dennis J.S., Horst I., Howe C.J., Lea-Smith D.J., Smith, A.G. (2010) Biodiesel from algae: challenges and prospects. Current Opinion in Biotechnology. 21(3):277-86

Sevenier, R. et al. (1998) High level fructan accumulation in a transgenic sugar beet. Nat. Biotechnol. 16, 843–846

Shewmaker, C.K. et al. (1999) Seed-specific overexpression of phytoene synthase: increase in carotenoids and other metabolic effects. Plant J. 20, 401– 412

Shintani, D. & DellaPenna, D. (1998) Elevating the vitamin E content of plants through metabolic engineering. Science 282, 2098–2100

Smeekens, S. (1997) Engineering plant metabolism. Trends Plant Sci. 2, 286–287

Society of Toxicology (2003) The Safety of Genetically Modified Foods Produced through Biotechnology. Toxicological Sciences 71, 2-8

Sottosanto, J.B., Saranga Y., Blumwald E. (2007) Impact of AtNHX1, a vacuolar Na+/H+ antiporter, upon gene expression during short and long term salt stress in Arabidopsis thaliana. BMC Plant Biology. 7:18

Sprenger, N. et al. (1997) Fructan synthesis in transgenic tobacco and chicory plants expressing barley sucrose: fructan 6-fructosyltransferase. FEBS Lett. 400, 355–358

Stark-Lorenzen, P.N.B. et al. (1997) Transfer of a grapevine stilbene synthase gene to rice (Oryza sativa L.). Plant Cell Rep. 16, 668–673

Stearidonic Acid (SDA) 2011, Monsanto Omega-3 Soybeans, http://www.monsanto.com/products/Pages/sda-omega-3-soybeans.aspx)

Szankowski, I. et al. (2003) Transformation of apple (Malus domestica Borkh.) with the stilbene synthase gene from grapevine (Vitis vinifera L.) and a PGIP gene from kiwi (Actinidia deliciosa). Plant Cell Rep. 22, 141–149

Uauy, C. et al. (2006) Gene regulating senescence improves grain protein, zinc, and iron content in wheat. Science 314, 1298–1301

UPI United Press International (2002) Wheat Inhibits Colon Cancer

Ursin, V.M. (2003) Modification of plant lipids for human health: development of functional land-based omega-3 fatty acids. J. Nutr. 133, 4271–4274

USDA APHIS (2012) http://www.aphis.usda.gov/biotechnology/petitions_table_pending.shtml#not_reg

Vistive Gold Soybeans, http://www.monsanto.com/products/Pages/vistive-gold-soybeans.aspx

White, C.L.T.L. et al. (2001) Increased efficiency of wool growth and live weight gain in Merino sheep fed transgenic lupin seed containing sunflower albumin. J. Sci. Food Agric. 81, 147–154

Yang, S.H. et al. (2002) Expression of a synthetic porcine alpha-lactalbumin gene in the kernels of transgenic maize. Transgenic Res. 11, 11–20

Ye, X. et al. (2000) Engineering the provitamin A (beta-carotene) biosynthetic pathway into (carotenoid-free) rice endosperm. Science 287, 303– 305

Young, T.E. et al. (2004) Senescence-induced expression of cytokinin reverses pistil abortion during maize flower development. Plant J. 38, 910–922

Yu, J. & Ao, G. (1997) Expression of 10 kDa sulfur-rich prolamin gene of rice in potato. Acta Bot. Sin. 39, 329–334

Yu, O. et al. (2000) Production of the isoflavones genistein and daidzein in nonlegume dicot and monocot tissues. Plant Physiol. 124, 781–794

Yu, O. et al. (2003) Metabolic engineering to increase isoflavone biosynthesis in soybean seed. Phytochemistry 63, 753–763

Zeh, M. et al. (2001) Antisense inhibition of threonine synthase leads to high methionine content in transgenic potato plants. Plant Physiol. 127, 792–802

Zhang, P., Wang W.Q., Zhang G.L., Kaminek M., Dobrev P., Xu J., Gruissem W. (2010) Senescence-inducible expression of isopentenyl transferase extends leaf life, increases drought stress resistance and alters cytokinin metabolism in cassava. J. Integr. Plant Biol.52, 653–669

Zhao, Z.Y.G.K. et al. (2002) Nutritionally improved transgenic sorghum. Plant Biotechnology 2002 and Beyond. Proceedings of the 10th IAPTC&B Congress, Orlando, FL, June 23–28, 2002

BOARD OF SCIENTIFIC AND POLICY ADVISORS

BOARD OF SCIENTIFIC AND POLICY ADVISORS (CONTINUED)

BOARD OF SCIENTIFIC AND POLICY ADVISORS (CONTINUED)

BOARD OF SCIENTIFIC AND POLICY ADVISORS (CONTINUED)

BOARD OF SCIENTIFIC AND POLICY ADVISORS (CONTINUED)

BOARD OF SCIENTIFIC AND POLICY ADVISORS (CONTINUED)